"Healthful eating is never about taste. All diets taste good, or no one would follow them. If the food wasn't tasty, no matter how "good for you" it was, you wouldn't eat it for more than a meal or two. Taste is a non-factor. But if it were about taste, The High Energy Diet has to rank among the tastiest, if not the most delicious, of all diets."

~

"Fruit has been our natural food throughout history. It is sweet and juicy, two of the most desired features any food could offer. We make poor facsimiles of fruit, at best, when we offer our loved ones cookies and milk, coffee and cake, pie and ice cream, hot chocolate and marshmallow, or other such replicas of the real thing. Fruit is the real deal."

The New High Energy Diet *Recipe Guide*

by
Dr. Douglas N. Graham

The New High Energy Diet Recipe Guide
by Dr. Douglas N. Graham

Published by: **FoodnSport Press**

609 N. Jade Drive

Key Largo, FL 33037

U.S.A.

All rights reserved. No part of this book may be reproduced or transmitted in any form or by any means, electronic or mechanical, including photocopying, recording or by any information storage and retrieval system without written permission from the author, except for the inclusion of brief quotations for review or reference purposes. Inquiries regarding requests to reprint all or part of this book may be sent to *foodnsport@aol.com*

Copyright © 1998, 2007 by Douglas N. Graham

This book is printed in China.

Logo and cover design by: **Janie Gardener**

Photography by: **Carina Honga, Amanda Williams and Lennie Mowris**

Layout by: **Lennie Mowris**

ISBN: 978-1-893931-25-4
1-893831-25-6

Dedication

This book is dedicated to two super achievers—my best teachers, my loving parents, Marty and Bea. Your constant love, encouragement, and support have allowed me to grow into the person that I am today. I have attempted to learn as much as possible from you, to incorporate the best from both of you into my personality, and to learn from your mistakes.

I hope that I can surpass your highest dreams for me, and that the recipes in this book will bring you great enjoyment and long-lasting health.

Also by Dr. Douglas N. Graham

Grain Damage
Nutrition and Athletic Performance
The Perpetual Health Calendar
The 80/10/10 Diet

For information about Dr. Graham's books, CDs, DVDs, articles, lecture series, Health & Fitness Weeks, supervised fasting, and other educational products or events, visit *www.foodnsport.com*.

Table of Contents

Foreword	8
Preface	10
Preface for the New Edition	11
Advantages of a Raw-Food Diet	12
Conversion Charts	14
Balancing Your Meals	17
10 Tips for Transitioning to Raw	18
Acknowledgements	19
Fruit Meals	23
Soups	35
Fruit Soups	36
Fruit and Vegetable Soups	40
Vegetable Soups	50
Salads & Slaws	65
Fruit Salads	66
Super Salads	76
Cold Slaws	94
Toppings & Dressings	103
Celebration Food	127
Appetizers	128
Pies	136
Cookies, Candies and Other Delightful Desserts	146
Index	164

Foreword

If they'd needed a cookbook in Eden, this could have been the one, full of heavenly delicacies from the orchard and garden. This is euphoria food' too-good-to-be-good-for-you food; succulent, flavorful, exquisite food. The fact that it also has the ability to ensure and even restore a high level of health and vitality is because this is not so much a cookbook as a cook-less book: Not one of the glorious recipes on its pages requires a stove, oven, or microwave, just a working relationship with your local greengrocer.

Adding more fresh, uncooked foods to your diet is a treat that can also be a challenge. We need help in trading in old habits for better ones, and the beautiful, easy-to-prepare recipes on these pages provide that help deliciously.

Most of us were brought up on heavy foods, cooked and overcooked, with animal products on center stage and refined sweets and snack foods playing major supporting roles. We're better educated today and are planning meals with lower cholesterol and lower overall fat levels, more fiber, and fewer unnecessary chemical additives. These are major steps in the direction of lifetime health and nonstop energy. In The High Energy Diet Recipe Guide however, Dr. Graham takes us even further, to a way of eating that supplies us with all the vital force of natural foods, straight from the earth to the table.

As a longtime health writer, I have learned about numerous dietary philosophies. I have seen that any change from the burger-and-fries norm brings about noticeable improvement in the life of the person who tries it.

For this reason, I applaud those who are sharing with us ways to cut the fat in our cooking methods, those who encourage a gradual shift to a more vegetarian diet, the macrobiotic community, and those in medicine and dietetics who challenge the traditional Four Food Groups and are providing the public with better choices.

The New High Energy Diet Recipe Guide

The choice offered by Dr. Graham is, in my opinion, optimal. I base this not only on research and interviews with experts over the years, but on my own experience. I always eat "well", that is, I am a complete vegetarian; I do not eat fried foods or sugary, salty snack foods; I don't drink coffee or colas. Nevertheless, there are times when my vitality lags, when I don't feel infused with life first thing in the morning. Those are the times I need to remind myself to increase the fresh, uncooked fruits, vegetables and juices in my diet. When I do this, I bounce back in just a day or two—the way a droopy, dry houseplant perks up after a good watering.

Dr. Graham's recipes are just what the doctor ordered for keeping the level of fresh, unprocessed foods high in my diet and yours. Colorful fruits and vegetables are bursting with vitamins, minerals, and enzymes that cooking destroys in short order. The raw nuts and seeds used moderately in some of the recipes provide satiety and essential fatty acids without resorting to extracted oils or heated fats that are believed to have carcinogenic properties.

In addition to these physical benefits, featuring uncooked foods as the predominant part of your diet offers more subtle rewards as well: you'll spend less time in the kitchen; you'll have more hours for exercise, relaxation, expanding your mind, enjoying your hobbies, and being with those you love.

Be gentle with yourself while you embark upon this new adventure. It isn't all or nothing. It is instead a process for growth: try at least one of these recipes every day for a month. See what happens. See how you feel. Listen to your body. Trust it. Go with the seasons as they change. And remember that life produces life: the more living foods you include in your daily menus, the more living you are apt to pack into every day. And thanks to Dr. Graham, you can do this eating "fudge" and "chocolate pudding."

Victorian Moran

Author of *The Love-Powered Diet, Lit From Within, Creating a Charmed Life,* and *Younger By the Day*

Preface For The New Edition

Preface

Welcome to *The High Energy Diet Recipe Guide*. Using diet in the sense that what you eat is your diet, *The High Energy Diet Recipe Guide* will provide you with ideas for the ultimate in healthy eating. Whether you have been a long time purist or this is your first experiment with improving your food regime, *The High Energy Diet Recipe Guide* will help you to make great strides.

The outstanding feature of this book is that all the recipes are prepared from 100 percent raw foods. This is not to say that I think cooking is to be wholly avoided, more that I feel the reader is already competent in the art of creating cooked food dishes. This effort is geared towards helping you create healthy raw food recipes, which I hope will eventually become the mainstay of your diet. As you add in more uncooked meals, you will become fluent with the art of developing them on your own. Your overall percentage of uncooked foods will increase. I am striving to help you achieve a diet characterized by an abundance of raw foods in which fruits and vegetables predominate. This type of diet is universally accepted as one which will help you to reach your fullest health potential while doing yourself and the environment the least possible harm.

I have tried to be as specific as possible with this guide. However, I encourage you to experiment, substitute, and find out exactly what best suits your needs and satisfies your tastes. There is always room for questions. I appreciate your questions, comments, and feedback. Please let me know if you found this booklet useful, or how you think I could improve upon it. It is my sincerest wish that you find this book and its audio and video companions of immense value.

Preface for the New Edition

I first wrote *The High Energy Diet Recipe* Guide back in the late '80's, after countless requests from clients and patients for a book they could "live from." It was the first of its kind in that it was the first recipe book to use only whole, raw ingredients while utilizing proper food combining principles. To this day, it still remains the only book to have done so.

The New High Energy Diet Recipe Guide is an improved version in many ways. It includes more recipes, caloronutrient information for each recipe, and even a thought on healthful eating with each turn of the page.

Over the years, I have several times thought of phasing out *The High Energy Diet Recipe Guide*. Each time I read through it, however, I discovered that the recipes were still as valuable as they were when first written. I still get letters from people who use the book as their kitchen bible, and I still receive unsolicited email from people who have healed themselves from a wide variety of illnesses simply by following the recipes found within.

I hope you and your family enjoy *The New High Energy Diet Recipe Guide*. I trust that it will prove useful to you for many years to come. Hopefully, you will even improve on some of the recipes and will tell me about it when you do. Also, please feel free to send me pictures of your creations if you are ever so moved.

In health abundance naturally,
Dr Douglas N. Graham

Advantages of a Raw-Food Diet

The concept of this book is simple: to help you incorporate more uncooked recipes into your yearly meal plan. With practice you will become talented in serving raw-food dishes. As you increase your creativity and ability to make them delicious for you and your loved ones, you will find that eating garden-fresh food is really most satisfying.

The most obvious advantage of a raw-food diet is that every meal is **quick to prepare**. That is not to say that you can't prepare complex and fascinating meals, but rather you have the chance for each meal to be zero prep. Think of the time you usually spend in the kitchen preparing your average dinner. Compare that to a meal of watermelon. Bacon and eggs, pancakes, even a simple sandwich, all take preparation time that makes fruit look like the original "fast food".

To ensure that you get the absolute maximum in **nutritional density**, your foods must be eaten in a raw state with as little preparation as possible. Heat is one of the main destroyers of nutrients in your food. Enzymatic destruction begins at 116 degrees; vitamins lose potency at 130 degrees; and proteins become denatured above 161 degrees Fahrenheit. (Denatured proteins are unusable to your body, cannot be "re-natured," and have been linked with many diseases, including arthritis, heart disease and even cancer.) Try sticking your hand in boiling water for even a moment if you want to see proteins denatured. Even steaming vegetables requires a temperature of 212 degrees. While heated foods will deliver caloric density more easily than their raw counterparts, most of their nutrients have been cooked out.

If you are looking for calories without nutrients, cooked foods are the best source. If you want the highest possible nutrient count per calorie, eat raw foods.

Perhaps the most noticeable feature of a raw meal is that it is so incredibly **easy to clean up**. Greases are almost never present, and nothing is ever baked onto the pan or dish. Usually all it takes to clean up is a quick rinse with water only. Sometimes a sponge is necessary to remove a spot of nut butter or a sticky bit of fruit.

When you eat raw foods you can guarantee that your meals will always be **fresh**. Did you ever hear of someone giving themselves food poisoning on fresh fruit? Of course not, because when the food is raw, it is easy to tell if it has gone "bad." Fresh, whole, raw fruits and vegetables are rich in flavors, colors and textures too.

Perhaps the most important reason for eating a predominantly raw food diet is that it is **environmentally responsible**. No packaging need be manufactured or thrown away. Energy is not required to heat the food. Rain forests are saved rather than cut to raise livestock. Animals are nurtured, not destroyed. With more than 50 percent of our nation's water supply and 80 percent of our grains going to raise livestock, a diet high in fruits and vegetables is environmentally sound.

I sincerely hope that you will find the recipes in this book as delicious and easy to prepare as I do. I look forward to receiving your comments and suggestions. Enjoy your meals.

In Abundant Health,

Dr. Douglas N. Graham

Conversion Charts

We have listed the ingredients in the menu plan in terms of ounces and pounds, in order to provide an accurate caloronutrient breakdown. If you do not have a scale at home, the charts below can help you measure out the quantities called for. Eventually you will become proficient in estimating weights and average caloric content of various foods, a skill that will pay dividends in saved kitchen time.

Portion Equivalents: Vegetables (1 lb)	
Bell peppers	3 cups chopped, 4 medium
Broccoli	5 cups chopped, 0.75 bunch
Butter leaf lettuce	2.75 cups chopped
Cabbage	5 cups chopped, 0.5 med. head
Cauliflower	4.5 cups chopped, 1 head
Celery	4.5 cups chopped, 1 head
Cucumbers, peeled	3 cups sliced, 2.5 medium
Green/red leaf lettuce	12.5 cups shredded, 1 large head
Romain lettuce	9.5 cups shredded, 1 large head
Spinach	15 cups, 1 bunch
Tomatoes	2.5 cups chopped, 3.5 medium
Tomatoes, cherry	3 cups

The New High Energy Diet Recipe Guide

Portion Equivalents: Sweet Fruits (1 lb)	
Apples	4 cups sliced, 3.5 medium
Apricots	2.75 cups sliced, 13 medium
Bananas	3 cups sliced, 4 medium
Blackberries	3 cups
Blueberries	3 cups
Cantaloupe Melon	2.75 cups cubed
Casaba Melon	2.5 cups cubed
Cherries, sweet	4 cups with pits, 64 medium
Dates	12.5 cups pitted, 19 medjool
Figs	9 medium
Grapefruits	2 cups sectioned, 2 medium
Grapes	3 cups
Honeydew Melon	2.75 cups diced
Kiwis	2.5 cups, 6 medium
Mangos	2.75 cups sliced, 2 medium
Nectarines	2 cups sliced, 3 medium
Oranges	2.5 cups sectioned, 3.5 medium
Papayas	3 cups cubed, 1.5 medium
Peaches	2.7 cups sliced, 4.5 medium
Pears	2.75 cups sliced, 2.5 medium
Persimmons	2.75 medium
Pineapples	3 cups, 1 medium
Plums	2.75 cups sliced, 7 medium
Raisins	3 cups
Raspberries	3.5 cups
Strawberries	2.5 cups sliced, 38 medium
Tangerines	2 cups sections, 2 medium
Watermelons	3.75 cups, 0.5 of a large melon

Conversion Charts

Portion Equivalents: Overt Fats (specified below)	
Avocado (6–7 ounces)	1 medium
Almonds (1oz.)	23 kernels
Hemp seeds (1oz.)	4 tablespoons
Macadamia nuts (1 oz.)	10–12 kernels
Pecans (1 oz.)	20 halves
Pine nuts (1 oz)	140 pine nuts
Pistachios (1 oz.)	49 kernels
Sesame seeds (1 oz.)	3.5 tablespoons
Sunflower seeds (1 oz.)	5 tablespoons
Tahini (1 oz.)	2 tablespoons
Walnuts (1 oz.)	14 halves

Balancing Your Meals

A healthful diet can essentially be described as one rich in raw fruits and vegetables and low in dietary fat. However, the ideal diet is one where roughly 80% of our total calories are derived from carbohydrate sources such as whole, fresh, ripe fruit. Consuming an 80% carbohydrate diet automatically limits our consumption of fat and protein to only 10% of total calories each.

This recommendation is far from the diet the average person consumes. The average American consumes about 40% of their total calories from fat, regardless of omnivorous, vegetarian or vegan status. Whereas, the average person on a raw food diet consumes roughly 65-80% of their calories from fat through the liberal use of nuts, seeds, oils, coconut and avocados.

To help keep your fat percentage low, I suggest only eating fatty foods once per day and not everyday. Typically, fats would be consumed at your dinner meal, likely as a part of a large, luscious salad.

Switching our focus to consuming whole fresh fruits to fuel our daily energy needs may take practice, because few people are accustomed to eating the volume of food necessary to sustain our caloric needs through fresh fruit. Fruit is so high in fiber and water that it takes more physical space than it's water and fiber deficient cooked counterparts. In time as we begin to practice eating more volume we find our stomach readily accommodates this change in our habits. Please understand I am not suggesting over-eating—eating until we hurt—but rather exercising the natural elasticity of our stomachs by small increments each meal until a sustainable quantity of food can be consumed.

10 Tips for Transitioning to Raw

Fad diets are all the rage these days. In a world full of change, it can be unnerving to make bold lifestyle choices with confidence, knowing that the chance of you sticking with any nutritional program permanently is slim to none.

Thankfully, a diet composed of fresh, ripe, raw organic plant foods is not only delicious, but satisfying. The variety of foods available over the course of a year, and the health and freedom that come with eating this diet, will keep you smiling at each meal and enjoying every bite for a lifetime. To ensure that you reach this level of gratification, you want to be sure to start off on the right foot. To make this easier, here are 10 tips for ensuring your first taste of our natural diet gets you hooked.

1. Learn about fruit
2. Buy fresh, ripe, raw organic plant foods
3. Replace one meal of the day at a time with a raw fruit meal, starting with breakfast
4. Make friends with your local produce supplier
5. Develop a regular fitness program
6. Phase out all stimulants
7. Keep your fat intake below 10%
8. Seek simplicity at each meal
9. Seek variety over the course of the year
10. Enjoy each and every bite!

Acknowledgements

Helping people to achieve their goals brings me tremendous satisfaction. Many people should feel satisfied about my completing this book, without whose support it could never have been done.

To every staff member who shielded me and allowed me to sit and write saving me from interruption and distraction, thank you. Every hygienic professional, those who led before me as well as my colleagues, have motivated me with their unfailing examples of service and achievement.

Thank you, Shari, for your understanding. And for realizing that producing this book was, for me, a mission. Thank you for handling virtually everything else while this project held my full attention.

To my beloved wife and amazing baby girl. Rozi and Faychesca, you brighten every corner of my life. Thank you for all the years of love you have shown me. I simply don't know what I would do without you both.

To Laurie Masters, Gail Davis, and Christina Chadney whose exceptional editing skills made this book possible and legible.

To our photographers. Carina, thank you, for the time you put into the start of this project and for the delicious photographs. Amanda, thank you for coming in at the last minute to make this book even better. Your pictures are truly worth a thousand words.

Special thanks goes to Lennie Mowris. Lennie, it appears nothing happens without you. No words fully express my appreciation for all that you do. But through our friendship and teamwork I am certain you know what they are.

To everyone who has encouraged me to finish this project: patients, family, and friends. If hundreds of you had not demanded it repeatedly, I never would have written it.

THE RECIPES

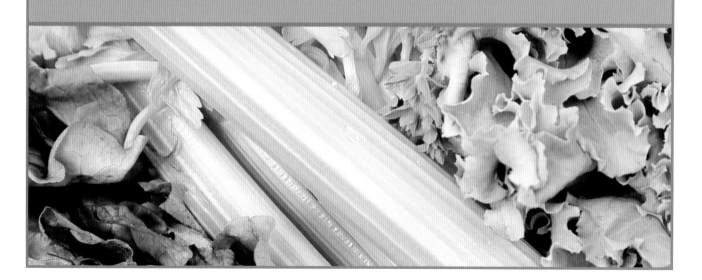

Fruit Meals

Everybody has a sweet tooth. In nature the desire for sweeter foods caused us to reach for fruits to satisfy our need for fuel and the vitamin C complex. Fruits are the best source of almost every vitamin. When people ask, "But don't you like sweets?" they are showing that they don't understand how many sweets you really do eat.

The mono fruit meal is the foundation to a healthful raw diet. The simplicity of these meals allows for the experience of tasting all the distinctions between separate pieces of fruit within the same variety and the nuances between individual varieties. The Earth gave us a tremendous gift when she gave us the shapes, sizes and flavors fruit can be found. Learning to enjoy fruit as it comes isn't a difficult task for most people.

Almost any fruit is easily consumed as a meal unto itself, provided you have sufficient quantity to meet your caloric needs. There is no limit to the amount of fruit you are allotted in a day, so eat until you are satisfied!

Fruit Meals

Tropical Treats

Be on the lookout for foods you have not tried. Whereas simplicity is instrumental to the good digestion of each meal, diversity in your overall diet is essential to ensure nutritional sufficiency. Eating new foods usually results in unexpected pleasures. Try items you haven't yet tasted ask your grocer for assistance. Here are a few of my tropical favorites.

Custard's First Stand

There is a class of fruit called anona, eaten worldwide in warm climates. These include atemoya, cherimoya, anon, sugar apple, sour sop, custard apple, pond apple, rolinea, and many more.

The Mona Lisa Custard Apple is increasingly being cultivated in South Florida. If you can find it, buy it. Allow it to fully ripen and enjoy it with almost any other fruit.

I'm in love with Rolinea!

Rolinea—one of the anona's—is a fruit which, when fully ripe, has the color, texture, consistency, and taste of rice pudding. Enjoy it all by itself for a delightful meal.

Guanabana

Fresh Guanabana (also called soursop) is a sub-acid-to-sweet treat eaten throughout the tropics.

How's your mom been?

The red mombin plum tree may be the most prolific of all fruit-bearing trees. It yields up to six crops per year, each time literally covering the tree with small, fire-engine-red, delicious plums. The taste is something like pineapple and peach. Enjoy them fresh, or put frozen mombin in a bowl of fresh-squeezed orange juice.

Jamaican Cherry Crush

The Jamaican cherry tree, possibly the fastest-growing woody fruit tree, produces a prodigious quantity of fruit the year round. A small white fruit with tiny internal seeds, the Jamaican cherry taste reminds some of caramel, others of the smell of hot buttered popcorn. Mash a pint of cherries into a bowl with 1 or 2 fully ripe bananas. Delicious.

Durian

The Durian is native to Southeast Asia and is often referred to as the "king of fruits." This fruit has a smell and spiny appearance that many may find offensive, especially initially. Frequently, people find that after the first taste the smell of this fruit shifts from an onion-like scent to a sweeter smell. Durian's texture is reminiscent of custard. After just a few bites most people find they can't stop eating it!

Fruit Meals

Quarter Shake

2 oz. avocado
32 oz. fresh-squeezed orange juice
Directions: Blend.

Classic Sweet Shakes

Blend any of the following with banana:

water
date
raisins
persimmon
blueberry
pear
pineapple
apple
papaya
mango
vanilla
carob

† Feel free to experiment! Changing the amount of water, the number and/or kind of fruit, or simply the temperature of the ingredients all change the meal dramatically. Enjoy!

The calories in the chart to the right are for approximately 6-8 bananas with roughly 8-16 oz. of other fruit depending on selection.

Classic Sweet Shake	Carb	Protein	Fat
Grams	277	11	4
Calories	997	39	31
% of total calories	93	4	3
Total calories for this course	1067		

† this chart is an approximation based on averages of the suggested combinations.

Have you tried a new fruit this year?

Quarter Shake	Carb	Protein	Fat
Grams	99	7	10
Calories	382	29	88
% of total calories	76	6	18
Total calories for this course	499		

Fruit Meals

Tropical Delight

32 oz. orange juice
12 oz. papaya
12 oz. pineapple
1 oz. young coconut meat

Directions: Blend fruit first, then add coconut.

Peachberry Shake

2 lbs. frozen peaches
1 lb. berry of your choice

Directions: Fill blender half full of water. Add peaches until almost full. Top off with berries.

Tropical Delight	Carb	Protein	Fat
Grams	119	8	11
Calories	446	29	95
% of total calories	78	5	17
Total calories for this course	570		

Fruit makes life sweeter.

Peachberry Shake	Carb	Protein	Fat
Grams	102	9	3
Calories	361	32	24
% of total calories	86	8	6
Total calories for this course	417		

Fruit Meals

Thick Shake

16 oz. fresh-squeezed orange juice
1 lb. frozen strawberries
2 oz. avocado

Directions: Blend strawberries in orange juice until smooth and then add avocado.

Five-Star Shake

16 oz. fresh-squeezed orange juice
3 oz. frozen five-star fruit (carambola)

Directions: Blend orange juice with carambola.

Thick Shake	Carb	Protein	Fat
Grams	87	7	11
Calories	324	27	89
% of total calories	74	6	20
Total calories for this course	440		

Nature does the work, we just put on the finishing touches. You are gonna love these dishes, in excellent health, for a lifetime.

Five-Star Shake	Carb	Protein	Fat
Grams	57	4	1
Calories	222	17	11
% of total calories	89	7	4
Total calories for this course	250		

Fruit Meals

Classic "Juicy" Shakes

Choose any of the following ingredients:

orange with	berry
	tomato
	plum
strawberry with	mango
	peach
	grapefruit
papaya with	nectarine
	tangerine
	passion fruit
pineapple with	pear
	apple
	cherry

Directions: Blend.

Classic "Juicy" Shake	Carb	Protein	Fat
Grams	119	8	2
Calories	426	27	17
% of total calories	90	6	4
Total calories for this course	471		

† this chart is an approximation based on averages of the suggested combinations.

For a sweet and juicy life, eat fruit.

Soups

A good soup is hard to beat. The variety of soups on the raw diet seems virtually endless. Changing even one ingredient in your favorite combination will yield an entirely different flavor and experience. Keep in mind that any soup can make an amazing dressing as well!

Fruit soups greatly broaden your scope of variety and presentation. They make a tasty first course to any meal and are great served as a complete meal. Many people find that they can consume fruit more easily as soup than in any other form. Enjoy these sweet soups on a hot summer night when you would rather be cool.

Fruits, fruit-vegetables such as bell peppers, tomatoes and cucumbers, along with tender greens make amazing additions to our diet. If you find that fruit goes down better with a little non-sweet food added then these soups are ideal for you!

Vegetables are the number one source of virtually every mineral. For this reason incorporating vast amounts of leafy matter into our diets is an absolute necessity. To help you achieve eating *enough* vegetables the addition of soups is an easy alternative. Enjoyable as a complete meal vegetable soups are also excellent for using up large amounts of many vegetables at once, preventing spoilage.

Soups

Fruit Soups

Strawberry Soup

2 lbs. fresh or frozen strawberries
1 oz. raw, soaked cashews
2-4 oz. water

Directions: Blend strawberries with a little water. Separately blend cashews and water, and gently pour mix in a swirl over strawberries. Serve with thinly sliced strawberries floating on top.

† Optional: Use macadamia or pine nut in place of cashew.

Grapefruit Soup

32 oz. fresh squeezed pink grapefruit juice
1 lb. berry (variety of your choice)

Directions: Pour berries into a bowl and drown in juice.

Grape Soup

16 oz. orange juice
8 oz. grapefruit juice
4 oz. grapes (variety of your choice)

Directions: Pour the citrus mixture over grapes.

† Optional: Use lychee, fresh or frozen, instead of grapes

Strawberry Soup	Carb	Protein	Fat
Grams	78	11	15
Calories	283	41	123
% of total calories	63	9	28
Total calories for this course	447		

Who needs condiments when you are eating fruit?

Grapefruit Soup	Carb	Protein	Fat
Grams	129	8	3
Calories	489	33	26
% of total calories	89	6	5
Total calories for this course	548		

Grape Soup	Carb	Protein	Fat
Grams	87	5	2
Calories	337	19	13
% of total calories	91	5	4
Total calories for this course	369		

Soups

Sweet Apple Soup

10 oz. Red Delicious apples
6 oz. raisins
2 to 4 cups water
6 oz. sapodilla

Directions: Chop 3 apples into chunks, and grate the remaining apple. Blend chopped apples with the other ingredients and stir in grated apple.

† If you can't find sapodilla, use pear instead and ¼ teaspoon cinnamon. If this may be too sweet for you, blend in ½ stalk celery for every apple to reduce the sweetness.

Melon Soup

Any melon,
or any combination of melons

† For a variation try adding fresh mint leaves!

Directions: Put a little melon into a blender and blend briefly. As soon as it liquefies, add more. Keep adding melon and blending until you have a blender full. Serve thick or strain for a delicious summer treat.

Sweet Apple Soup	Carb	Protein	Fat
Grams	502	7	3
Calories	738	24	24
% of total calories	94	3	3
Total calories for this course	786		

How much fruit is enough? All you care for.

Melon Soup (per 32 oz serving)	Carb	Protein	Fat
Grams	77	8	2
Calories	279	29	15
% of total calories	86	9	5
Total calories for this course	323		

Soups

Fruit and Vegetable Soups

Strawberry-Cucumber Soup

1 lb. strawberries
8 oz. cucumbers

Directions: Peel the cucumber for smoother blending. Chop the cucumber and strawberries into a blender, reserving a little of each for a garnish. Blend, pour into a bowl and garnish.

Variation: Sweet Cukes

8 oz. mangoes
8 oz. cucumbers

Directions: Blend cucumbers and mango. Thinly slice cucumber and stir into blend. Pour and enjoy.

† Optional: Use lime juice and cilantro for a tasty treat.

Pineapple-Red Pepper Soup

1 lb. pineapple
½ lb. red bell peppers
¼ lb. tomato

Directions: Peel and core the pineapple. Core the red pepper. Blend the pineapple and red pepper. Dice the tomato and stir into the soup or leave it on top as a garnish.

Strawberry Cucumber Soup	Carb	Protein	Fat
Grams	40	4	2
Calories	160	16	18
% of total calories	82	8	9
Total calories for this course	194		

Fruit: A food for all seasons.

Variation: Sweet Cukes	Carb	Protein	Fat
Grams	40	2	1
Calories	160	8	9
% of total calories	90	5	5
Total calories for this course	177		

Pineapple-Red Pepper Soup	Carb	Protein	Fat
Grams	73	6	2
Calories	292	24	18
% of total calories	87	7	5
Total calories for this course	334		

Soups

Peach Heirloom Tomato Soup

8 oz. peaches
8 oz. heirloom tomatoes

Directions: Blend 6 oz. peaches with 6 oz. tomatoes. Thinly slice the remaining quarters of peach and tomato and stir them into the soup for added texture and color.

Kiwi Cucumber Soup

1 lb. kiwi
8 oz. cucumbers
2 oz. pomegranate seeds

Directions: Peel the kiwi and the cucumber. Blend 9 oz. kiwi and all the cucumber to form the soup base. Slice the remaining half of kiwi and stir it into the soup or arrange on top for garnish. Sprinkle pomegranate seeds on top for a splash of flavor and color.

Orange Verde Soup

8 oz. romaine lettuce
12 oz. Valencia oranges

Directions: Blend lettuce and ¾ of the oranges. Break apart the left over orange half into thin slices or wedges for garnish.

The New High Energy Diet Recipe Guide

Peach Heirloom Tomato Soup	Carb	Protein	Fat
Grams	32	4	1
Calories	128	16	9
% of total calories	84	10	6
Total calories for this course	153		

Have you discovered your "favorite" fruit?

Kiwi Cucumber Soup	Carb	Protein	Fat
Grams	81	7	3
Calories	324	28	27
% of total calories	85	7	7
Total calories for this course	379		

Orange Verde Soup	Carb	Protein	Fat
Grams	48	6	2
Calories	192	61	18
% of total calories	71	23	7
Total calories for this course	271		

Soups

Cabbage Tomato Soup

8 oz. fresh-squeezed orange juice
8 oz. of cabbage
4 oz. lettuce
4 oz. of tomatoes

Directions: Blend orange juice, cabbage, lettuce. Place puree into large serving bowl. Dice the tomato into chunks and stir into soup or sprinkle on top for a garnish.

Sweet Tomatoes

8 oz. mangoes
8 oz. heirloom tomatoes

Directions: Blend ¾ of the mango and ¾ of the tomato. Cut the remaining mango and tomato into small chunks and stir into soup blend.

Cabbage Tomato Soup	Carb	Protein	Fat
Grams	47	7	1
Calories	188	28	9
% of total calories	84	12	4
Total calories for this course	225		

Fruit blended with green is not a new idea. Newcomers call it a green smoothie, but for decades it has simply been a sweetened blended salad.

Sweet Tomatoes	Carb	Protein	Fat
Grams	49	3	1
Calories	196	12	9
% of total calories	90	6	4
Total calories for this course	217		

Soups

Papaya Gazpacho

1 lb. papayas
8 oz. tomatoes
2 oz. fresh basil

Directions: Deseed the papaya and cut it into small chunks. You can also blend it if you prefer. Dice the tomato and finely chop the basil. Add them both to the papaya mixture. Enjoy!

Mango Fennel Soup

1 lb. mangoes
1 large sprig of fennel

Directions: Blend 12 oz. mangoes and the bottom 3/4 of the fennel sprig together. Pour into a bowl. Cut the remaining mango into small chunks and mix into the soup. Garnish with the top of the fennel sprig. Delicious!

The New High Energy Diet Recipe Guide

Papaya Gazpacho	Carb	Protein	Fat
Grams	57	6	2
Calories	228	24	18
% of total calories	84	9	7
Total calories for this course	270		

Wouldn't you prefer fruit? Peachy idea.

Mango Fennel Soup	Carb	Protein	Fat
Grams	77	2	1
Calories	270	7	8
% of total calories	95	2	3
Total calories for this course	285		

Soups

Savory Sunrise Soup

8 oz. celery
8 oz. red bell peppers
8 oz. Valencia oranges

Directions: Blend all the celery, oranges and 6 oz. red pepper. Use the remaining red pepper for a garnish.

Variation: with Spinach

8 oz. spinach
8 oz. red bell peppers
8 oz. orange

Directions: Blend all of the ingredients and pour into a bowl.

Savory Sunrise Soup	Carb	Protein	Fat
Grams	47	6	2
Calories	188	24	18
% of total calories	82	10	8
Total calories for this course	230		

Got fruit? Got vegetables? Got all we need.

Variation: with Spinach	Carb	Protein	Fat
Grams	49	11	2
Calories	196	44	18
% of total calories	76	17	7
Total calories for this course	258		

Soups

Vegetable Soups

Classic Tomato Celery Soup

1 ¼ lbs. tomatoes
12 oz. stalks celery
juice of 1 lemon

Directions: Use the 'S' blade of a food processor to liquefy tomatoes. Cut the celery into 2-inch lengths and add them to the tomatoes. Process briefly until the celery mixes in, but leave it slightly coarse. Serve with lemon on the side.

Borscht

12 oz. beets
1 lb. celery
milk of 1 young coconut
2 oz. young coconut meat
8 oz. water

Directions: Peel, shred, and blend the beets and celery. Separately, blend the coconut milk and meat of coconut with sufficient water to make "cream." Serve in individual bowls and add "cream" separately.

Classic Tomato Celery Soup	Carb	Protein	Fat
Grams	36	8	2
Calories	125	25	13
% of total calories	76	16	168
Total calories for this course	161		

Humans have eaten raw, organic fruit throughout their entire history.

Borscht	Carb	Protein	Fat
Grams	60	12	16
Calories	222	43	134
% of total calories	55	11	34
Total calories for this course	399		

Soups

Tomato Basil Soup

1 lb. of tomatoes
5 sun-dried tomato halves
fresh basil to taste

Directions: Soak sun-dried tomatoes for ten minutes. Blend three tomatoes, basil and sun-dried tomatoes together. Pour into a bowl. Chop remaining tomato into chunks and place in the center of the soup. Garnish with one fresh basil leaf.

Celery-Red Pepper Soup

8 oz. celery
8 oz. red bell peppers
8 oz. tomatoes

Directions: Blend the celery and red pepper together to make the soup base. Dice the tomato and add to the top.

Tomato Basil Soup	Carb	Protein	Fat
Grams	29	6	2
Calories	116	24	18
% of total calories	73	15	11
Total calories for this course	158		

Vegetables yield the lowest calories per bite of any food.

Celery Red Pepper Soup	Carb	Protein	Fat
Grams	31	2	6
Calories	124	18	24
% of total calories	75	11	14
Total calories for this course	166		

Soups

Creamy Gazpacho Soup

2 lbs. tomatoes
1.5 oz cashews (or cashew butter)
4 oz. red bell peppers
12 oz. celery
juice of 1 lime

Directions: Blend tomatoes and cashews, reserving one tomato. Dice the bell peppers, celery and remaining tomato. Pour soup over diced veggies and garnish with lime rounds.

† Optional: Use macadamia or pine nut in place of cashew

Orange Pepper & Cuke Soup

½ lb. orange peppers
½ lb. cucumbers
4 oz. strawberries

Directions: Blend the orange pepper and the cucumber together. Slice the strawberries and use for garnish, color and texture.

Creamy Gazpacho Soup	Carb	Protein	Fat
Grams	71	20	21
Calories	253	69	171
% of total calories	51	14	35
Total calories for this course	493		

Vegetables: High in nutrition, excellent flavor and low in calories. What more could you ask for?

Orange Pepper & Cuke Soup	Carb	Protein	Fat
Grams	28	4	1
Calories	112	16	9
% of total calories	82	12	6
Total calories for this course	137		

Soups

Creamy Tomato Celery Soup

1 ¾ lbs. tomatoes
10 oz. celery
1.5 oz. cashews
8 oz. water
4 oz. pineapple

Directions: Blend tomatoes, celery and cashew with water. Garnish with finely chopped pineapple.

† Optional: Use macadamia or pine nut in place of cashew.

Creamy Corn Soup

8 oz. field-fresh corn
6 oz. cucumbers
4 oz. water
6 oz. avocado
4 oz. green cabbage
4 oz. sprouts
sprig of parsley

Directions: Blend corn, cucumbers, avocado and water. Pour over finely chopped cabbage and chopped sprouts. Garnish with parsley sprig.

Creamy Tomato Celery Soup	Carb	Protein	Fat
Grams	67	71	21
Calories	241	62	169
% of total calories	51	13	36
Total calories for this course	472		

Fruit: real food for health-lovers.

Creamy Corn Soup	Carb	Protein	Fat
Grams	72	18	29
Calories	255	63	230
% of total calories	47	11	42
Total calories for this course	548		

Soups

Cuke Soup

1 ½ lbs. cucumber
1 oz tahini
6 oz. celery
8 oz. cabbage

Directions: Blend cucumbers, tahini, celery, and water as desired. Pour over shredded cabbage.

Cauliflower Soup

1 lb. cauliflower
1 oz. pecans
12 oz. water
1 lb. broccoli
3 oz. daikon radish
4 oz. red bell peppers
8 oz. zucchini

Directions: Blend cauliflower, pecans, and water. Finely chop the broccoli, daikon, bell pepper and zucchini. Mix all ingredients together and garnish around the edges with cauliflower and broccoli greens.

Cuke Soup	Carb	Protein	Fat
Grams	40	14	15
Calories	145	50	126
% of total calories	45	16	39
Total calories for this course	321		

Many seeds of fruits are arranged in a five-pointed star. It is no wonder there are so many five-star fruit meals.

Cauliflower Soup	Carb	Protein	Fat
Grams	76	29	23
Calories	263	100	182
% of total calories	49	18	33
Total calories for this course	545		

Soups

Gazpacho

1 lb. tomatoes
juice of 1 lime
4 oz. celery
4 oz. cucumbers
4 oz. red bell peppers
4 oz. orange bell peppers
4 oz. yellow summer squash
1 tbsp. sunflower seeds

Directions: Blend tomatoes and lime-juice. Chop remaining vegetables. Pour blend over chopped vegetables. Garnish lightly with sunflower seeds. Serve slightly chilled.

Gazpacho II

1 ½ lbs. tomatoes
1 lb. celery
4 oz. alfalfa sprouts

Directions: Blend tomatoes and celery. Pour over finely chopped alfalfa sprouts.

† Optional: Add 1 diced avocado if desired.

Gazpacho	Carb	Protein	Fat
Grams	49	16	16
Calories	174	55	129
% of total calories	49	15	36
Total calories for this course	358		

You want variety? In the course of a year over 200 varieties of fruits and 50 vegetables are typically sold in most grocery stores.

Gazpacho II	Carb	Protein	Fat
Grams	44	14	3
Calories	151	46	22
% of total calories	69	21	10
Total calories for this course	219		

Soups

Asparagus Soup

8 oz. celery
4 oz. alfalfa sprouts
1 oz. raw, soaked, blanched almonds
1 lb. asparagus
    (without tips)
8 oz. water
6 oz. yellow pepper

Directions: Chop celery and sprouts. Crush almonds. Blend asparagus with water, and pour over chopped vegetables and crushed almonds. Garnish with thinly sliced yellow pepper and the asparagus tips.

Asparagus Soup	Carb	Protein	Fat
Grams	45	24	16
Calories	155	83	128
% of total calories	42	23	35
Total calories for this course	366		

*Fruits and Vegetables:
the all-time health foods*

Salads & Slaws

Salads come in as many shapes sizes and colors as there are fruits and vegetable on the planet! We can make fruit-only salads, fruit and greens salads or strictly vegetable salads. They are excellent when joined with a delicious soup or delicious as a meal of their own.

In addition to the bounty of leafy green salads you can shred a vegetable and have a slaw. Slaws are one of my favorite vegetable meals, because they are easy to eat and have potential for so much variety. The mechanical digestion has been partially done by machines making slaws a good choice for people with chewing difficulties or for people who are not yet used to the amount of chewing necessary for a fresh, vegetable meal.

Almost every evening salad or slaw makes up a big part of, or sometimes our entire, dinner meal. I have found that there is much more room for variety, night after night, if I keep down the number of ingredients in each salad. These are some of my tried-and-true favorites, but in reality I make a different salad almost every night.

† Keep in mind the fat content of each of your salads, and the averaging of fat over the period of a day, week or month.

Salads & Slaws

Fruit Salads

Tropical Lunch

1 lb. carambola (star fruit)
12 oz. freshly squeezed grapefruit juice
6 oz. tangerines
6 oz. lettuce

Directions: Dice carambola into a bowl and cover with grapefruit juice. Decorate with sliced tangerine. Serve with lettuce.

Citrus Ambrosia

1 lb. oranges
6 oz. grapefruit
8 oz. pineapple
4 oz. tangerines
4 oz. star fruit
6 oz. young coconut milk
1 oz. young coconut meat

Directions: Dice the fruit and grate the coconut. Mix all ingredients with the coconut milk.

Tropical Lunch	Carb	Protein	Fat
Grams	90	10	3
Calories	332	36	24
% of total calories	85	9	6
Total calories for this course	392		

Next time you find yourself in front of a camera, smile and say "lychees!"

Citrus Ambrosia	Carb	Protein	Fat
Grams	135	10	12
Calories	491	38	96
% of total calories	79	6	15
Total calories for this course	625		

Salads & Slaws

Ambrosia

8 oz. grapes
8 oz. tart apples
6 oz. young coconut milk
1 oz. young coconut meat

Directions: Blend grapes and grate apples. Mix with the coconut milk and meat. Serve in a pie dish and top with sliced stone fruit.

Strawberrry-Mango Bliss

1 ¼ lbs. mangoes
12 oz. strawberries

Directions: Peel, core, and dice mangos. Toss with slivers of strawberries.

Ambrosia	Carb	Protein	Fat
Grams	78	4	11
Calories	288	15	90
% of total calories	73	4	23
Total calories for this course	393		

Is the peach the "northern mango" or is the mango the "southern peach?" Either way, you can substitute one for the other in almost any recipe.

Strawberry-Mango Bliss	Carb	Protein	Fat
Grams	123	5	3
Calories	437	19	21
% of total calories	92	4	4
Total calories for this course	477		

Salads & Slaws

Jam "Handwich"

12 oz. banana
6 oz. romaine lettuce leaves
8 oz. "Your Basic Jam" (p.102)

Directions: Slice bananas lengthwise. Place on lettuce leaves. Spread banana with Your Basic Jam to taste. Roll up the lettuce and serve as a "handwich."

Fruit Compote

8 oz. black or blond figs
water as needed
6 oz. apples
6 oz. celery

Directions: Cut off the stems and blend figs with enough water to make a runny sauce. Shred apples and celery. Mix all ingredients together and serve.

Jam "Handwich"	Carb	Protein	Fat
Grams	152	7	2
Calories	548	26	15
% of total calories	93	4	3
Total calories for this course	589		

Fruit: Pure solar energy

Fruit Compote	Carb	Protein	Fat
Grams	172	9	3
Calories	616	33	21
% of total calories	92	5	3
Total calories for this course	670		

Salads & Slaws

Sweet Grapes

12 oz. green grapes
12 oz. bosc pears
8 oz. blackberries

Directions: Chop pears into bite sized pieces and toss with grapes and berries.

Kiwi Orange Crush

1 lb. kiwi
1 lb. mandarin orange

Directions: Peel kiwi and oranges. Separate oranges into sections. Crush 4 oz. of kiwi into a bowl. Chop remaining kiwi into thin half moons. Toss with crushed kiwi. Enjoy!

Sweet Grapes	Carb	Protein	Fat
Grams	133	7	3
Calories	477	24	22
% of total calories	91	5	4
Total calories for this course	523		

Fruit: Our perfect food nutritionally.

Kiwi Orange Crush	Carb	Protein	Fat
Grams	127	9	4
Calories	455	32	30
% of total calories	88	6	6
Total calories for this course	517		

Salads & Slaws

Mango Delight

1 ¾ lbs. mangoes
juice of 1 lime

Directions: Dice mangoes into a bowl and squeeze on the lime-juice. This simple, delicious meal is tough to surpass.

Blue Cheeks

1 lb. peaches
12 oz. blueberries

Directions: Slice peaches into 1-inch chunks. Layer blueberries overtop.

Mango Delight	Carb	Protein	Fat
Grams	140	4	2
Calories	498	15	17
% of total calories	94	3	3
Total calories for this course	530		

Fruit is easier to digest than any other food.

Blue Cheeks	Carb	Protein	Fat
Grams	93	7	2
Calories	329	24	18
% of total calories	89	6	5
Total calories for this course	371		

Salads & Slaws

Super Salads

Classic Tomato Salad

1 lb. tomatoes
6 oz. avocado
1 lb. cucumbers
4 oz. leaf lettuce

Directions: Dice tomatoes and avocado. Toss well, mound on a bed of leaf lettuce and place thinly sliced cucumber discs around tomato and avocado.

Corn Salad

1 ¼ lbs. fresh corn
12 oz. cucumbers
12 oz. celery
1 oz. parsley
add herbs or dried vegetables to taste

Directions: Cut the kernels off the ears of corn. Dice cucumbers and celery, and toss with finely chopped parsley. Use herbs or dried vegetables sparingly to taste.

Classic Tomato Salad	Carb	Protein	Fat
Grams	52	13	27
Calories	185	42	212
% of total calories	42	10	48
Total calories for this course	439		

Shall we have dinner? Lettuce have salad.

Corn Salad	Carb	Protein	Fat
Grams	127	23	8
Calories	441	82	63
% of total calories	75	14	11
Total calories for this course	586		

Salads & Slaws

Italian Salad, Southern Style

1 lb. broccoli
8 oz. cauliflower
12 oz. romaine lettuce
4 oz. whole green olives
1 oz. sun-dried tomatoes
8 oz. water
1 oz. leaves fresh basil

Directions: Break the broccoli and cauliflower into pieces. Chop romaine lettuce and olives. Re-hydrate tomatoes in the water, and then blend with basil into a dressing.

Greek Salad

8 oz. tomatoes
6 oz. zucchini
8 oz. mung bean sprouts (or other non-starchy sprout)
6 oz. red bell peppers
3 oz. young okra
2 oz. black olives
2-4 oz. water

Directions: Dice the tomatoes and zucchini. Finely chop the red pepper, and chop the okra. Mix cut ingredients into the mung bean sprouts. Blend olives and water into a dressing.

† Recommended: Rinse and soak olives for several hours, changing the water hourly, to remove brine.

Italian Salad, Southern Style	Carb	Protein	Fat
Grams	75	27	21
Calories	256	94	164
% of total calories	50	18	32
Total calories for this course	514		

Young tender greens are the ideal vegetables.

Greek Salad	Carb	Protein	Fat
Grams	48	15	8
Calories	162	50	60
% of total calories	60	18	22
Total calories for this course	272		

Salads & Slaws

Crushed Berry Salad

8 oz. baby spinach
4 oz. tomatoes
4 oz. cucumbers
4 oz. blackberries
4 oz. raspberries
4 oz. peaches

Directions: Place the mixed greens in a bowl. Peel the cucumber if desired. Slice the tomato and cucumber and mix with the salad greens. Pour the berries into a separate bowl and mash with a fork. Cut the peach into small pieces and mix with the berries. Pour over the salad.

Tomato Heaven!

1 lb. tomatoes
9 oz. baby greens
2 oz. hemp seeds

Directions: Place the mixed greens into a bowl. Cut the tomatoes into wedges. Mix tomatoes and hemp seeds thoroughly into a bowl. Dress the salad with the tomato mix.

Crushed Berry Salad	Carb	Protein	Fat
Grams	51	12	3
Calories	204	48	27
% of total calories	73	17	10
Total calories for this course	279		

If a salad seems unsatisfactory, you didn't have enough fruit beforehand.

Tomato Heaven!	Carb	Protein	Fat
Grams	40	20	11
Calories	160	80	99
% of total calories	47	24	29
Total calories for this course	339		

Salads & Slaws

Broccomole Salad

1 lb. broccoli
12 oz. red bell peppers
12 oz. green leaf lettuce
6 oz. avocado
juice of 1 key lime

Directions: Finely chop broccoli and bell peppers. Mix with coarsely chopped leaf lettuce and arrange on a platter. Blend avocado and lime-juice and pour over salad.

Blackberry-Sesame Salad

1 lb. romaine lettuce
4 oz. tomatoes
8 oz. blackberries
2 oz. raw, mechanically hulled tahini

Directions: Finely chop the lettuce into a large bowl. Slice the tomato into wedges and toss with lettuce. Blend blackberries and tahini and pour over top.

Broccomole Salad	Carb	Protein	Fat
Grams	81	25	28
Calories	278	85	220
% of total calories	47	15	38
Total calories for this course	583		

Can you name 10 varieties of lettuce?
Bibb, Buttercrunch, Boston, Cos, Romaine, Green leaf, Red leaf,
Iceburg, Little Gem, Lolla Rosa

Blackberry-Sesame Salad	Carb	Protein	Fat
Grams	45	16	17
Calories	180	64	153
% of total calories	45	16	39
Total calories for this course	397		

Salads & Slaws

Strawberry-Parsley

1 lb. red leaf lettuce
8 oz. cherry tomatoes
8 oz. strawberries
1 oz. parsley

Directions: Tear lettuce into a bowl. Slice the tomatoes in half and stir into the lettuce. Blend the strawberry and parsley together to make the dressing. Enjoy!

† Variation: Use fresh Fennel in place of parsley

Orange-Walnut Salad

8 oz. romaine lettuce
4 oz. oranges
1 oz. chopped walnuts

Directions: Chop the lettuce into a bowl. Peel the oranges and cut them into small pieces and place in a separate bowl. Stir the walnuts in with the oranges and put over the lettuce. Simply Delicious!

Strawberry-Parsley	Carb	Protein	Fat
Grams	40	10	3
Calories	160	40	27
% of total calories	70	18	12
Total calories for this course	227		

A sure sign of stimulant addiction is when the thought of a salad seems "blah."

Orange-Walnut Salad	Carb	Protein	Fat
Grams	25	8	20
Calories	100	32	180
% of total calories	32	10	58
Total calories for this course	312		

Salads & Slaws

Heirloom Avocado Salad

8 oz. romaine lettuce
8 oz. cucumbers
12 oz. tomatoes
6 oz. hass avocado
¼ cup cilantro

Directions: Chop lettuce into a bowl. Peel and slice cucumber and add it to the lettuce. Chop the cilantro. In a separate bowl, chop the tomato and avocado into chunks and stir together with the cilantro. When the tomato, avocado and cilantro have blended together pour the mixture over the lettuce and cucumber.

Layered Salad

1 lb. tomatoes
12 oz. romaine lettuce
6 oz. avocado
12 oz. cucumbers
6 oz. red bell peppers
8 oz. alfalfa sprouts

Directions: Layer slices of tomato, cucumber, avocado, red pepper (rings) and sprouts on a bed of lettuce leaves. Serve in a clear glass bowl.

Heirloom Avocado Salad	Carb	Protein	Fat
Grams	43	10	28
Calories	172	40	252
% of total calories	37	9	54
Total calories for this course	464		

I am a proud supporter of vegetable rights!

Layered Salad	Carb	Protein	Fat
Grams	70	24	29
Calories	244	85	233
% of total calories	44	15	44
Total calories for this course	562		

Salads & Slaws

Papaya Salad

1 lb. butter lettuce
1 lb. papayas
juice of 1 lime

Directions: In a bowl cut the papaya into small chunks. Mix it with the lime-juice until it all melds together. Use the lettuce leaves as wrappers and use your papaya-lime blend as the filling.

Strawberry-Almond Salad

1 lb. red leaf lettuce
4 oz. cucumbers
4 oz. strawberries
1 oz. almonds

Directions: Chop lettuce into a large bowl. Peel and slice the cucumber and mix with the lettuce. Blend the strawberries and almonds together and use it to dress the salad.

Papaya Salad	Carb	Protein	Fat
Grams	57	9	1
Calories	228	36	9
% of total calories	84	13	3
Total calories for this course	273		

Lettuce tastes fantastic if you're hungry.

Strawberry-Almond Salad	Carb	Protein	Fat
Grams	27	13	16
Calories	108	52	144
% of total calories	36	17	47
Total calories for this course	304		

Salads & Slaws

Green "Keys" Salad

12 oz. broccoli
8 oz. green leaf lettuce
12 oz. celery
6 oz. sprouts
12 oz. tomatoes
4 oz. young coconut milk
1 oz. young coconut meat

Directions: Chop all of the vegetable matter together. Mix with the blended milk and meat of the coconut.

MacCabbage Salad

1 lb. green cabbage
12 oz. romaine lettuce
12 oz. tomatoes
1 oz. raw macadamia nuts
3 to 4 oz. water
juice of 1 lime

Directions: Chop the cabbage and romaine. Mix in diced tomatoes. Dress with blend of macadamia nuts, water, and lime-juice.

Green "Keys" Salad	Carb	Protein	Fat
Grams	67	27	14
Calories	232	91	107
% of total calories	54	21	25
Total calories for this course	430		

Popeye was a good influence and created an entire generation of spinach eaters.

MacCabbage Salad	Carb	Protein	Fat
Grams	56	16	24
Calories	197	56	186
% of total calories	45	13	42
Total calories for this course	439		

Salads & Slaws

Eggless Salad

6 oz. yellow summer squash
6 oz. cucumbers
10 oz. celery
10 oz. green cabbage
4 oz. romaine lettuce

Directions: Grate the squash, cucumber, celery, and cabbage. Serve on a bed of romaine lettuce.

† Optional: Add whipped avocado, pine nuts, or diced white mild radish.

Rainbow Salad

4 oz. beets
4 oz. carrots
4 oz. yellow squash
4 oz. broccoli
4 oz. red cabbage
4 oz. sweet yellow corn
6 oz. avocado
8 oz. sweet yellow bell peppers

Directions: Grate the beets, carrots, squash, broccoli and cabbage. Use a round platter and place arcs of equal portions of various vegetables. Blend corn, avocado and bell pepper, an excellent "pot of gold" dressing.

Eggless Salad	Carb	Protein	Fat
Grams	37	10	2
Calories	127	35	13
% of total calories	73	20	7
Total calories for this course	175		

*Vegetables come in every color of the rainbow.
It is their color that indicates their predominant phytonutrients.*

Rainbow Salad	Carb	Protein	Fat
Grams	92	18	28
Calories	327	66	225
% of total calories	53	11	36
Total calories for this course	618		

Salads & Slaws

Cold Slaws

Cashew Slaw

1 lb. green cabbage
12 oz. red cabbage
8 oz. celery stalks
6 oz. red bell peppers
2 oz. cashew cream (p.103)

Directions: Grate the cabbages, celery, and red pepper. Mix thoroughly with the cashew cream.

† Optional: Sprinkle lightly with powdered caraway seed. Use macadamia or pine nut insted of cashew.

Red Slaw

12 oz. red cabbage
6 oz. cucumbers
1 ¼ lbs. tomatoes
1 oz. pecans

Directions: Shred the cabbage and cucumbers. Blend the tomatoes and pecans to make a dressing. Pour dressing over shredded vegetables. Mix and Serve.

Cashew Slaw	Carb	Protein	Fat
Grams	76	20	14
Calories	265	69	113
% of total calories	60	15	25
Total calories for this course	447		

Enjoy tougher veggies such as cabbage, broccoli, kale and cauliflower by finely shredding them into a slaw.

Red Slaw	Carb	Protein	Fat
Grams	55	13	22
Calories	196	48	180
% of total calories	47	11	42
Total calories for this course	424		

Salads & Slaws

Caul Slaw

1 lb. cauliflower
8 oz. celery
1.5 oz. blanched raw almonds, soaked
4 oz. water
12 oz. tomatoes

Directions: Shred the cauliflower and celery. Dice the tomatoes and set aside. For the dressing, blend nuts with water on high for one minute. Mix all the ingredients with the dressing, adding in the diced tomatoes.

† For a sweet dressing blend with orange juice instead of water, plus the juice of one lemon or lime if you prefer a "bite."

Sprout Slaw

8 oz. alfalfa sprouts
8 oz. mung bean sprouts
10 oz. green cabbage
6 oz. avocado
1 lb. tomatoes
4 oz celery

Directions: Chop the alfalfa sprouts into half-inch lengths. Mix in the mung bean sprouts and shredded cabbage. Blend avocados and tomatoes together and pour over the vegetables. Mix and serve garnished with whole celery stalks.

Caul Slaw	Carb	Protein	Fat
Grams	53	23	23
Calories	187	81	185
% of total calories	41	18	41
Total calories for this course	453		

Not sure if you're hungry?
Try eating celery. If it tastes great,
you are hungry.

Sprout Slaw	Carb	Protein	Fat
Grams	74	28	28
Calories	255	97	220
% of total calories	45	17	38
Total calories for this course	572		

Salads & Slaws

Tomato Fennel Slaw

8 oz. red cabbage
1 lb. tomatoes
1 frond of fennel (garnish and flavor)

Directions: Chop all ingredients and stir together into a bowl.

† If the flavor of fennel is not palatable to you than feel free to substitute any mild herb that you do enjoy such as basil or cilantro.

Celery Slaw

1 ¼ lbs. celery
8 oz. red bell peppers
12 oz. tomatoes
1.5 oz. walnuts
1 lemon

Directions: Grate the celery and bell peppers. Blend tomatoes and walnuts into a dressing. Mix it all together and serve with lemon on the side.

Tomato Fennel Slaw	Carb	Protein	Fat
Grams	38	7	2
Calories	152	28	18
% of total calories	77	14	9
Total calories for this course	198		

Vegetables all taste good on their own. They do not need to be mixed into a salad and dressed to be palatable.

Celery Slaw	Carb	Protein	Fat
Grams	54	16	30
Calories	192	56	241
% of total calories	39	11	50
Total calories for this course	489		

Salads & Slaws

Green Slaw

1 lb. green cabbage
6 oz. avocado
6 oz. broccoli
8 oz. water
4 oz. red bell peppers

Directions: Shred the cabbage. Blend the avocados, broccoli, and water for a dressing. Mix everything together. Garnish with thinly sliced red bell pepper.

Orange-Fennel Slaw

8 oz. cabbage
9 oz. oranges
1 shoot of fennel top

Directions: Finely chop the cabbage and fennel into a bowl. Chop the oranges into small pieces and stir into the cabbage-fennel salad.

Green Slaw	Carb	Protein	Fat
Grams	70	19	27
Calories	245	65	214
% of total calories	47	12	41
Total calories for this course	524		

Want to add volume to your dressings?
Add cabbage or some cucumber to stretch it.

Orange Fennel Slaw	Carb	Protein	Fat
Grams	70	8	1
Calories	280	32	9
% of total calories	87	10	3
Total calories for this course	321		

Toppings & Dressings

Toppings, the most visible portion of any dish, derive their names—sauces, dips, spreads, icings, fondues, jams—from their intended use and consistency. The same recipe can be used as a topping for fruit ice cream, as a spread over fruit compote, or as an icing for fruit pies.

Most of my sweet sauces consist of one dried fruit blended with water. I make them for sweetness as well as color. I am especially fond of pineapples, (pale yellow) persimmon (red), carob (dark brown), raisin (dark), date (light brown), canistel (deep yellow), and mamey (pink).

Salad dressings, the most common topping, are so easy I am repeatedly amazed at how few people make their own. Most dressings are composed of two main ingredients, something fatty and something with a bite. Traditionally, this means oil and vinegar. I use a nut, seed, or avocado for the oily component and any acid fruit for the bite. After that it is simply a matter of using available vegetables and herbs to create the flavor you strive for. Be brave and remember, "for success, dress."

Have fun making toppings. Get good at it, for these finishing touches are what your dining guests will remember most.

Toppings & Dressings

Toppings

Piña Colada Sauce

8 oz. pineapple
8 oz. oranges
4 oz. young coconut milk
1 oz. young coconut meat

Directions: Peel pinapple and oragnes. Deseed oranges if necessary. Combine with coconut milk and meat and blend.

Papaya Spread

12 oz. papaya
juice of 1 lime

Directions: Peel and de-seed the papaya; cut into 2-inch squares. Slowly add into blender, adding lime to taste. You may need a tiny bit of water depending on the papaya.

† Optional: Add 1 pint strawberry or ½ a pineapple

Hawaiian Crush

8 oz. pineapple
8 oz. fresh-squeezed orange juice
1.5 oz. nuts (variety of your choice)

Directions: Blend.

Piña Colada Sauce	Carb	Protein	Fat
Grams	66	5	10
Calories	239	18	85
% of total calories	70	5	25
Total calories for this course	342		

There are always sweet fruits coming into season. Think nature was trying to tell us something?

Papaya Spread	Carb	Protein	Fat
Grams	38	2	1
Calories	135	8	4
% of total calories	92	5	3
Total calories for this course	147		

Hawaiian Crush	Carb	Protein	Fat
Grams	65	11	19
Calories	243	40	163
% of total calories	54	9	37
Total calories for this course	446		

Toppings & Dressings

Orange and Walnut Sauce

1.5 oz. raw walnuts
3 oz. freshly-squeezed orange juice

Directions: Soak walnuts in water for 12 to 24 hours, changing the water several times or else the sauce will be quite tart. Blend walnuts and orange juice on high for 1 minute.

Apple-Walnut Grape Spread

1 oz. raw, soaked walnuts
4 oz. apples
4 oz. concord
(or other dark grapes)

Directions: Blend the nuts, grate the apples, and crush the grapes. Mix lightly.

Orange and Walnut Sauce	Carb	Protein	Fat
Grams	15	7	28
Calories	55	26	235
% of total calories	17	8	75
Total calories for this course	316		

Fresh fruit served at the beginning of the meal is always a welcomed treat.

Apple Walnut Grape Spread	Carb	Protein	Fat
Grams	38	5	19
Calories	139	20	157
% of total calories	44	6	50
Total calories for this course	316		

Toppings & Dressings

Your Basic Jam

8 oz. dried fruit
8 oz. water

Directions: Soak fruit in water for 8 hours. Blend to desired consistency by adding a bit more water or a bit more dried fruit.

Cashew Cream

1.5 oz. raw cashews, whole or pieces

Directions: Soak cashews for 4-24 hours in enough distilled water to cover completely. Blend nuts and water on high until smooth.

† Optional: Use macadamia or pine nut in place of cashew.

Salsa

8 oz. tomatoes
4 oz. tomatillos
4 leaves basil
2 leaves oregano
8 oz. cucumbers
4 oz. red bell peppers
4 oz. mild Daikon radish
1 lb. vegetables (variety of your choice)

Directions: In a blender or food processor, mix the tomatoes, tomatillos, basil, and oregano. Mince cucumber, bell pepper, and Daikon radish. Pour mixed ingredients over minced ingredients. Serve with vegetables.

The New High Energy Diet Recipe Guide

Your Basic Jam	Carb	Protein	Fat
Grams	145	6	1
Calories	522	20	9
% of total calories	94	4	2
Total calories for this course	551		

Talk about versatility! Every topping, dressing, dip or spread can be use interchangeably.

Cashew Cream	Carb	Protein	Fat
Grams	17	10	25
Calories	64	39	211
% of total calories	20	12	68
Total calories for this course	314		

Salsa	Carb	Protein	Fat
Grams	57	15	4
Calories	194	51	32
% of total calories	70	18	12
Total calories for this course	277		

Toppings & Dressings

Any Nutty Icing

1.5 oz. any raw, soaked nuts

Directions: Place soaked nuts in the blender. Add water slowly and blend to desired consistency.

† Note: Nuts go in the blender first, but too many nuts may burn out your blender. Start with a small amount and a bit of water. Add more nuts and water. You're more likely to run out of nuts before you run out of water, so put the nuts in first and add water as needed.

Avocado Icing

8 oz. avocado

Directions: Blend until creamy, adding water if necessary.

† Optional: blend in two stalks of celery

Any Nutty Icing	Carb	Protein	Fat
Grams	7	6	26
Calories	27	24	219
% of total calories	10	9	81
Total calories for this course	270		

Icing on the cake can be as quick as an avocado in the blender.

Avocado Icing	Carb	Protein	Fat
Grams	19	5	33
Calories	71	7	275
% of total calories	20	5	75
Total calories for this course	363		

Toppings & Dressings

Dressings

Sweet Russian Dressing

4 oz. tomatoes
2 oz. red bell peppers
6 oz. avocado
juice of 1 lime
2 oz. celery

Directions: Blend tomatoes, bell pepper, avocado and lime-juice. Mix in finely diced celery.

Cashew Cucumber Dressing

4 oz. cucumbers
1 oz. raw, soaked cashews

Directions: Peel cucumbers and blend with cashews to desired consistency.

† Optional: Use macadamia or pine nut in place of cashew.

Cool Satsuma Dressing

4 oz. satsuma tangerines
4 oz. cucumbers

Directions: Simply blend!

Sweet Russian Dressing	Carb	Protein	Fat
Grams	26	5	25
Calories	96	20	206
% of total calories	30	6	64
Total calories for this course	322		

Be sure to dress for salad success!

Cashew Cucumber Dressing	Carb	Protein	Fat
Grams	11	6	13
Calories	41	22	107
% of total calories	24	13	63
Total calories for this course	170		

Cool Satsuma Dressing	Carb	Protein	Fat
Grams	18	2	1
Calories	64	6	4
% of total calories	87	8	5
Total calories for this course	74		

Toppings & Dressings

Mango-Red Pepper Dressing

4 oz. mangoes
4 oz. red bell peppers

Directions: Peel and de-seed the mango. Blend and Pour!

Variation: Pineapple Rouge Dressing

4 oz. pineapple
4 oz. red peppers

Directions: Peel pineapple and de-seed red pepper. Blend.

Sunflower Vegetable Dressing

1.5 oz. sunflower seeds
2-4 oz. water
4 oz. of a mix of: broccoli, cauliflower, carrots, celery, and/or bell peppers work well

Directions: Blend all of the ingredients together.

The New High Energy Diet Recipe Guide

Mango-Red Pepper Dressing	Carb	Protein	Fat
Grams	26	2	1
Calories	92	6	5
% of total calories	89	6	5
Total calories for this course	103		

Master making salad dressings and you will be the honored chef at every meal.

Pineapple Rouge Dressing	Carb	Protein	Fat
Grams	21	2	1
Calories	74	6	4
% of total calories	88	7	5
Total calories for this course	84		

Sunflower Vegetable Dressing	Carb	Protein	Fat
Grams	13	12	21
Calories	49	43	178
% of total calories	18	16	66
Total calories for this course	270		

Toppings & Dressings

Tahini Dressing

1 oz. raw, sesame tahini
2 oz. water
juice of 1 lemon

Directions: Blend. Add water until creamy.

† Buy raw sesame tahini at your health food store or make it by running raw, hulled sesame seeds in the food processor with as little water as possible.

Mango-Raspberry Dressing

4 oz. mangoes
4 oz. raspberries

Directions: Peel mangoes and cut away seed. Blend with rapsberries for a tasty treat!

Tahini Dressing	Carb	Protein	Fat
Grams	11	5	14
Calories	42	19	112
% of total calories	24	11	65
Total calories for this course	173		

***Need a salad dressing in a hurry?
Blend celery with any berry and you're done!***

Mango Raspberry Dressing	Carb	Protein	Fat
Grams	33	2	1
Calories	118	7	8
% of total calories	89	5	6
Total calories for this course	133		

Toppings & Dressings

Coco Loco Dressing

1.5 oz. young coconut meat
3 oz. young coconut milk
3 oz. pineapple

Directions: Blend coconut matter and add pineapple to reach the desired consistency.

Sunny's Favorite Dressing

1.5 oz. raw sunflower seeds
6 oz. acid fruit (variety of your choice)

Directions: Run through food processor until creamy.

Pistachio Cucumber Dressing

4 oz. cucumbers
1 oz. pistachios

Directions: Blend until smooth.

Coco Loco Dressing	Carb	Protein	Fat
Grams	20	2	15
Calories	76	9	123
% of total calories	37	4	59
Total calories for this course	208		

Many salads require no dressing at all. Consider a salad of tomato, cucumber and parsley.

Sunny's Favorite Dressing	Carb	Protein	Fat
Grams	29	11	21
Calories	108	41	177
% of total calories	33	13	54
Total calories for this course	326		

Pistachio Cucumber Dressing	Carb	Protein	Fat
Grams	10	7	13
Calories	39	24	109
% of total calories	23	14	63
Total calories for this course	172		

Toppings & Dressings

Open Sesame Dressing

1.5 oz. raw sesame tahini
5 oz. lime-juice
1 oz. fresh mint
water as needed

Directions: Blend, adding sufficient water to create a creamy dressing.

† Optional: Add one herb (parsley, oregano, basil, rosemary, cilantro, or arugula) to completely change this dressing.

Pineapple-Macadamia Dressing

6 oz. pineapple
1.5 oz. raw macadamia nuts

Directions: Peel and core pineapple. Cut into chunks and blend first. Add in the nuts.

Open Sesame Dressing	Carb	Protein	Fat
Grams	25	9	21
Calories	91	33	166
% of total calories	31	11	58
Total calories for this course	290		

Make two different salads.
Blend one and pour it over the other.

Pineapple Macadamia Dressing	Carb	Protein	Fat
Grams	27	4	32
Calories	101	16	270
% of total calories	26	4	70
Total calories for this course	387		

Toppings & Dressings

Herb Dressing

12 oz. tomatoes
1 oz. of one of sprig of: anise, basil, oregano, or rosemary
6 oz. celery

Directions: Blend tomatoes and herb, adding celery as needed to thicken.

† Optional: Add in the juice of one lemon.

Creamy Herb Dressing

6 oz. tomatoes
6 oz. avocado
1 oz. of one of: arugula, dill, cilantro, or purslane

Directions: Blend.

† Optional: For a saltier flavor add celery to taste.

Apricot Celery Dressing

2 oz. celery
8 oz. apricots

Directions: Blend and Enjoy.

Herb Dressing	Carb	Protein	Fat
Grams	20	5	1
Calories	67	17	9
% of total calories	72	18	10
Total calories for this course	93		

The simple addition of a leafy herb can make an old dressing a new favorite.

Creamy Herb Dressing	Carb	Protein	Fat
Grams	22	6	25
Calories	81	20	208
% of total calories	26	6	68
Total calories for this course	309		

Apricot Celery Dressing	Carb	Protein	Fat
Grams	27	4	1
Calories	96	13	8
% of total calories	82	11	7
Total calories for this course	117		

Toppings & Dressings

Tomato Olive Dressing

6 oz. tomatoes

1.5 oz. pitted, whole black olives

Directions: Briefly blend.

Orange Juice Special Dressing

1 oz. raw cashews

4 oz. fresh-squeezed orange juice

Directions: Run cashews through a food processor until fine. Add orange juice to reach the desired consistency.

Sweet Tomato Walnut Dressing

4 oz. tomatoes

1.5 oz. raw walnuts

2 oz. freshly-squeezed orange juice

Directions: Blend. Add orange juice to reach desired consistency.

The New High Energy Diet Recipe Guide

Tomato Olive Dressing	Carb	Protein	Fat
Grams	9	2	5
Calories	33	7	40
% of total calories	41	9	50
Total calories for this course	80		

An acid fruit and a fatty food is all it takes to have a winning dressing.

Orange Juice Special Dressing	Carb	Protein	Fat
Grams	20	6	13
Calories	77	23	108
% of total calories	37	11	52
Total calories for this course	208		

Sweet Tomato Walnut Dressing	Carb	Protein	Fat
Grams	16	8	28
Calories	60	29	235
% of total calories	19	9	72
Total calories for this course	324		

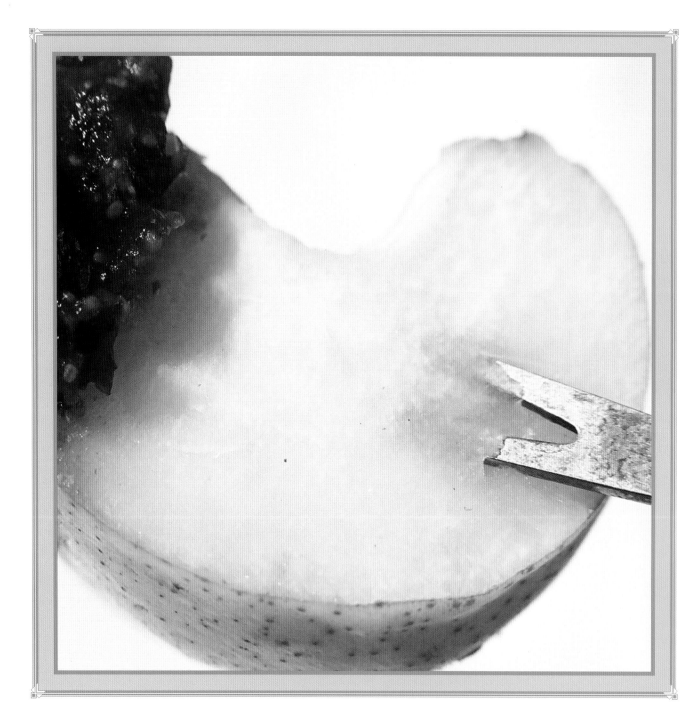

Celebration Food

Who said eating simple fare needed to be boring? Anything but. The following section includes appetizers, pies and desserts for all your special occasions, dinner guests, and holidays. During these times, the dishes which are heavier, richer, and tend to look more traditional seem more appropriate. Even though the proportions of the recipes are for more than one person, unlike the rest of the book, the nutritional information provided is adjusted for one serving, so you can eat your pie and count it too!

The appetizers in this book only scratch the surface of all the varieties of ways we can create healthful, simple meal starters. Much of the art in making tasty treats is the presentation. A beautifully designed fruit or vegetable platter is always more than welcome at any party.

When it comes to healthy pies and desserts anything goes! Imgination is the only limitation. Pies can be wholesome, delicious, and beautiful. Your fruit pies will be talked about, and copied, every time you let your loved ones try them.

†Keep in mind it is easy to overeat on these sweet meals so take your time. When eating, include lettuce and/or celery with your sweet meals. Balance your sugar intake with your activities, and remember to always listen to your body.

Celebration Food

Appetizers

Fruit Fondue

12 oz. bananas
12 oz. mango
12 oz. apples
12 oz. pears
12 oz. celery or lettuce

Directions: Dice equal amounts of each fruit. Arrange on a platter with toothpicks in each piece, or with fondue forks. Serve lettuce or celery alongside.

4 oz. dates
4 oz. raisins
4 oz. dried pineapple
water as needed

Directions: Blend each and put into separate bowls. Makes 3 thick sauces. Serves 8

Nutty Fondue

12 oz. acid fruit of your choice
1 lb. vegetables of your choice

Directions: Arrange on a platter.

2 oz. cashews (or choose macadamia or pine nut)
2 oz. almonds
2 oz. sesame seeds
water as needed

Directions: Blend each nut or seed with water to make three different dips. Serve in classic fondue style, with fondue forks, along with the platter of vegetables and "acid" fruit. Serves 8

Fruit Fondue	Carb	Protein	Fat
Grams	28	13	7
Calories	99	44	85
% of total calories	43	19	38
Total calories for this course	227		

*Slice a tasty tomato and serve it,
your creative genius will be appreciated by all.*

Nutty Fondue	Carb	Protein	Fat
Grams	13	5	10
Calories	48	20	86
% of total calories	32	13	55
Total calories for this course	154		

Celebration Food

Vegetable Fondue

8 oz. broccoli
8 oz. cauliflower
8 oz. celery
8 oz. yellow bell peppers
8 oz. zucchini

Directions: Cut equal amounts of each vegetable and arrange them on a platter.

Variation 1:
4 oz. tomatoes
4 oz. tomatillos
4 oz. celery
dried beet powder

Variation 2:
8 oz. cucumbers
basil
oregano
8 oz. avocado

Variation 3:
4 oz. cabbage
4 oz. corn
4 oz. red bell peppers
rosemary

Directions: Using a food processor create 3 thick fondue sauces. Serve with fondue forks or as finger food. Serves 8

Vegetable Fondue	Carb	Protein	Fat
Grams	16	4	5
Calories	57	14	39
% of total calories	54	13	33
Total calories for this course	110		

Veggies and fruit sound odd together?
Bananas wrapped in a lettuce leaf is a true taste treat.
Try it, you'll be surprised.

Celebration Food

Veggie Nut Hors D'Oeuvres

4 oz. soaked nuts (variety of your choice)
12 oz. celery stalks
1 ¼ lbs. broccoli florettes
12 oz. red bell peppers
12 oz. summer squash

Directions: Finely grind the soaked nuts, and finely cut broccoli florettes. Puree celery and bell peppers. Serve on slices of squash.

† Add other vegetables to satisfy your taste.
Serves 6

Cuke Cashew Sandwiches

4 oz. raw, soaked cashews
8 oz. cucumbers
8 oz. red bell peppers
1 lb. celery or cabbage

Directions: Blend nuts with cucumber and bell peppers. Spread on celery stalks or cabbage leaves.
Serves 6

† Optional: Use macadamia or pine nut in place of cashew.

The New High Energy Diet Recipe Guide

Veggie Nut Hors D'Oeuvres	Carb	Protein	Fat
Grams	28	13	7
Calories	99	44	85
% of total calories	43	19	38
Total calories for this course	227		

Appease your appetite with fruit!

Cuke Cashew Sandwiches	Carb	Protein	Fat
Grams	12	5	9
Calories	44	18	71
% of total calories	33	13	54
Total calories for this course	133		

Celebration Food

Coconut Dip With Pear and Apple

2 oz. young coconut meat
6 oz. young coconut milk
12 oz. pears
1 ¾ lbs. apples
8 oz. celery

Directions: Finely chop the celery. Use the "S" blade of a food processor to mix all ingredients.

Serves 6

Almond Dip with Pineapple & Strawberries

2 oz. raw, soaked blanched almonds
1 lb. pineapple
12 oz. strawberries

Directions: Blend.

Serves 6

Coconut Dip With Pear and Apple	Carb	Protein	Fat
Grams	31	1	4
Calories	112	5	29
% of total calories	77	3	20
Total calories for this course	146		

There are few appetizers more fun than fondue.

Almond Dip With Pineapple and Strawberries	Carb	Protein	Fat
Grams	16	2	5
Calories	58	11	41
% of total calories	52	10	38
Total calories for this course	109		

Celebration Food

Pies

Sweet Mango Pie
1 lb. dates
1 ¾ lbs. mango
de-seeded and chopped
5 oz. sliced banana

Directions: Soak dates for one hour. Homogenize them in a food processor or Champion Juice with the blank plate. Spread dates evenly onto a pie plate, up the sides and over the edges. Run frozen mango through a Champion juicer with the blank plate to make mango ice cream. Fill pie plate. Top decoratively with sliced banana. Freeze and serve.

Serves 8

Sweet Mango Pie	Carb	Protein	Fat
Grams	63	2	1
Calories	228	6	4
% of total calories	96	3	1
Total calories for this course	238		

Want a sweet treat?
Try some ripe fruit.

Celebration Food

Carobanana Mint Pie

1 lb. pitted dates
3 lbs. frozen banana
1 tbsp. raw carob powder
1 tbsp. diced mint leaves
8 mint sprigs

Directions: Press a layer of pitted dates into a glass pie dish. Run 24 oz. frozen bananas through a Champion juicer with the blank plate. Layer into the pie dish about ¾ inches thick. Put remaining bananas in a food processor with carob powder and diced mint leaves. Mix until bananas look completely dark and layer onto the pie. Decorate with mint sprigs spaced evenly around the circumference. Freeze and serve.

Serves 8.

All-American Pie

10 oz. frozen apples
14 oz. frozen strawberries
12 oz. frozen blueberries
3 oz. star fruit

*Directions: Use a Champion juicer with the blank plate to make **separate batches** of ice cream out of the apples, strawberries, and blueberries. On a rectangular platter place alternating rows of frozen apple and frozen strawberry to make stripes. Place the frozen blueberry in a square in the top left corner. Cut a fresh star fruit into ½-inch thick, star-shaped slices and place them on and around the blue area of the pie. Serves 8.*

Carobanana Mint Pie	Carb	Protein	Fat
Grams	63	2	1
Calories	225	7	3
% of total calories	93	3	1
Total calories for this course	235		

Have you eaten enough fruit today?

All-American Pie	Carb	Protein	Fat
Grams	30	2	1
Calories	109	6	6
% of total calories	90	5	5
Total calories for this course	121		

Celebration Food

Figgy Pie

12 oz. blond calmyrna figs
12 oz. black mission figs
1 ½ oz. Good Ol' Ice Cream (p.156)
5 oz. sliced bananas
6 oz. sliced peaches

Directions: Cut off the stems and soak figs. Use the food processor to mix black figs with just enough water to make a piecrust. Fill crust with "Good Ol' Ice Cream", one layer of banana, and one layer of peach. Make a top layer of blond figs. Freeze and Serve.

Serves 8

Figgy Pie	Carb	Protein	Fat
Grams	80	4	1
Calories	287	15	10
% of total calories	92	5	3
Total calories for this course	312		

Fruit is nature's dessert.

Celebration Food

Berry Good Pie

10 oz. frozen strawberries
12 oz. blueberries
8 oz. blackberries
4 oz. young coconut meat
6 oz. young coconut milk
6 oz. very ripe pineapple
4 oz. raspberries

Directions: Use a Champion juicer with the blank plate. Mix enough frozen strawberries to make a "crust." Add a layer of blueberries and a layer of blackberries. Top with a layer of thinly sliced strawberries. Blend coconut milk and meat with pineapple, and pour over pie dish until it is full. Decorate with a ring of raspberries. Freeze and serve.

Serves 8

Mango Pie

4 oz. dates
12 oz. pecans
1 ¼ lbs. strawberries
1 ¼ lbs. mango ice cream (see "Good 'Ol Ice Cream" p.156)

Directions: Soak pecans in water overnight. Use the "S" blade of a food processor to reduce nuts and dates to dough by slowly adding water. Spread dough into a pie plate up onto and over the sides. Add even layers of strawberry and then mango ice cream (made in a Champion juicer with the blank plate). Decorate the perimeter with whole strawberries turned upside down. Freeze and serve.

Serves 8

Berry Good Pie	Carb	Protein	Fat
Grams	20	2	5
Calories	72	7	43
% of total calories	60	5	35
Total calories for this course	122		

Dessert is the most satisfying part of any meal.

Mango Pie	Carb	Protein	Fat
Grams	34	5	31
Calories	126	18	258
% of total calories	31	5	64
Total calories for this course	402		

Celebration Food

Holiday Pie

8 oz. honey dates
4 oz. crushed walnuts
10 oz. bananas
8 oz. apples
5 oz. blueberries
6 oz. strawberries
6 oz. kiwi
4 oz fresh juice or blended fruit of choice

Directions: Use the "S" blade of a food processor to mix the dates and walnuts for a crust. Layer in sliced fruit for color and flavor. Pour juice or fruit blend over sliced fruit. Decorate with sliced fruit. The variations are endless. Use your imagination. Freeze and serve.

† Optional: Top with crushed coconut or nut butter dressing

† Note: This pie can be tough to digest. Serve with plenty of celery and lettuce.

Serves 8

Holiday Pie	Carb	Protein	Fat
Grams	44	4	10
Calories	160	14	80
% of total calories	64	5	31
Total calories for this course	254		

What's the best thing about High Energy recipes? They are as easy as pie.

Celebration Food

Cookies, Candies and Other Delightful Desserts

Dandy Candy

4 oz. any ground nut or seed
4 oz. any dried fruit
1 tsp. suitable spice
4 oz. freshly-squeezed fruit juice
4 oz. any nut butter

Directions: The variations on this basic recipe are endless. Mix in any of the ingredients above using equal portions of dry and moist ingredients to achieve a thick, gooey consistency. Shape as desired. Chill and serve.

† Make cookies softer by adding mashed ripe banana. † Decorate with currants, raisins, or coconut. Suitable spices include cinnamon, ginger, carob, vanilla, nutmeg, etc.

Serves 8

Dandy Candy	Carb	Protein	Fat
Grams	17	5	15
Calories	64	20	128
% of total calories	30	9	61
Total calories for this course	212		

Your love of sweets is right on the tip of your tongue.

Celebration Food

Could It Be Chocolate Pudding?

10 oz. dates
8 oz. black mission figs
1 quart water
1 tsp. raw carob powder

Directions: Blend dates and figs with water. More or less water may be needed depending on the 'dryness' of your fruit. Start with slightly less. Add carob powder, chill, and serve in pudding glasses.

Serves 6

Fudge

1 ½ lbs. bananas
1 lb. dates
2 inches vanilla bean
1 tsp. raw carob powder

Directions: Using the 'S' blade of a food processor, mix equal parts of date and banana. Process until completely smooth. You may need to add a few ounces of water, but add sparingly, as needed. Finely dice the vanilla and add to mixture. Keep food processor running and slowly add carob powder to taste. If mix is not dry enough or you would like a more heavy consistency, add a few ounces of chopped, dried fruit. Shape to desired form, freeze, and serve. This fudge will not freeze solid. Serve with lettuce and celery as a desert or as a sweet fruit meal.

Serves 8.

Could It Be Chocolate Pudding	Carb	Protein	Fat
Grams	60	2	1
Calories	216	8	4
% of total calories	95	3	2
Total calories for this course	228		

Every meal should be a dessert for the senses.

Fudge	Carb	Protein	Fat
Grams	63	2	1
Calories	224	7	3
% of total calories	96	3	1
Total calories for this course	234		

Celebration Food

Fudge S'Mores

1 lb. dates

6 oz. raisins

10 oz. "Fruit Gellee" (p. 152)

Directions: Soak dates and raisins in distilled water for 2 hours. Layer dates, raisins, and fruit Gellee into a square dish. Sprinkle lightly with raw carob powder. Freeze and serve with lettuce.

† Optional: Cover with crushed nuts.

† Note: This holiday treat can be a tough combination to digest.

Serves 8

Cookies

12 oz. dried fruit

8 oz. ripe bananas

4 oz. grated coconut

Directions: Add water slowly as needed. Form into balls, logs, or discs. Roll in grated coconut and serve frozen on a generous bed of lettuce.

† Optional: You may add minute amounts of minced ginger, raw carob powder, cocoa, or vanilla.

† Note: This can be a tough combination to digest. Enjoy some lettuce along with the cookies.

Serves 8

Fudge S'Mores	Carb	Protein	Fat
Grams	73	2	1
Calories	263	8	2
% of total calories	96	3	1
Total calories for this course	273		

After a meal of fruit, who needs dessert?

Cookies	Carb	Protein	Fat
Grams	37	2	10
Calories	136	9	77
% of total calories	61	4	35
Total calories for this course	222		

Celebration Food

Fruit Gelleé

1 lb. dried fruit
12 oz. water
6 oz. lettuce
6 oz. celery

Directions: Soak dried fruit in water. Use a food processor to mix the fruit in just enough water to mix. Allow to "set" for an hour or more in the refrigerator. If it does not set, add a few chopped raisins to soak up excess water. Serve with lettuce and celery.

Serves 8

Glazed Strawberries

2 oz. cold distilled water
4 oz. raw cashew butter
4 lbs. cold strawberries

Directions: Use the "S" blade of a food processor. Mix water and cashew butter until a thick creamy consistency is reached. Dip fresh cold strawberries into cashew mix. Serve with lettuce.

† Cashew butter can be purchased raw at your health food store.

Serves 8

The New High Energy Diet Recipe Guide

Fruit Geleé	Carb	Protein	Fat
Grams	38	2	1
Calories	135	7	3
% of total calories	93	5	2
Total calories for this course	145		

Eat dessert first... have some fruit.

Glazed Strawberries	Carb	Protein	Fat
Grams	22	4	7
Calories	79	15	57
% of total calories	52	10	38
Total calories for this course	151		

Celebration Food

Banana Split

1 ½. lbs. frozen bananas
1 ½ lbs. "Good Ol' Ice Cream" (p. 156)
1 lb. "Your Basic Jam" (p. 108)
6 fresh bananas

Directions: Run frozen bananas through a Champion juicer. Add 3 scoops ice cream to one "split" banana. Pour on "Your Basic Jam" of choice. Serves 6

Banana Ice Cream

3 lbs. frozen bananas

Directions: Peel and freeze bananas. Run uncut frozen bananas through a Champion juicer. This procedure will work in a food processor; you just have to go a little slower. It will even work in a blender; just go slowly and add a little water as needed. Serve immediately with lettuce.

Serves 8

Banana Split with Sweet Fruit Ice Cream	Carb	Protein	Fat
Grams	74	3	1
Calories	268	11	11
% of total calories	92	4	4
Total calories for this course	290		

Berries and cherries make everyone merry.

Banana Split with Juicy Fruit Ice Cream	Carb	Protein	Fat
Grams	67	3	1
Calories	241	10	6
% of total calories	94	4	2
Total calories for this course	257		

Banana Ice Cream	Carb	Protein	Fat
Grams	39	2	1
Calories	140	7	5
% of total calories	93	4	3
Total calories for this course	152		

Celebration Food

Chico Pudding

3 lbs. sapodilla

Directions: Skin and deseed sapodilla (chico) with a fork and serve.

† This fruit is worth seeking out. It can be ordered from fruit suppliers in south Florida. Fresh and raw, this fruit tastes similar to pears baked with cinnamon. Add over figgy puddin' or mashed banana for a real treat.

Serves 6

Good Ol' Ice Cream

4 lbs. frozen fruit, of your choice

Directions: Run frozen fruit through a Champion juicer using the blank plate. Banana and other sweet fruits come out like soft ice cream. Sub-acid fruits come out like sherbert. Acid fruits and melon have the texture of Italian ices. We prefer to eat the ice cream around the edges as it melts.

Serves 8

Figgy Puddin'

1 ½ lbs. dried figs (any variety)
6 oz. lettuce
6 oz. celery

Directions: Remove fig stems and soak overnight in enough water to cover them completely. Blend figs and their soak water. Add water to reach desired consistency. Serve with lettuce and celery.

Serves 6

Chico Pudding	Carb	Protein	Fat
Grams	45	1	3
Calories	164	4	20
% of total calories	87	2	11
Total calories for this course	188		

Everyone loves a sweet treat.

Good Ol' Ice Cream	Carb	Protein	Fat
Grams	34	1	1
Calories	121	4	4
% of total calories	94	3	3
Total calories for this course	129		

Figgy Puddin'	Carb	Protein	Fat
Grams	74	5	1
Calories	266	16	10
% of total calories	92	5	3
Total calories for this course	292		

Enjoy these dishes,

In excellent health,

Naturally,

Dr. Doug Graham

About the Author

Dr. Douglas Graham, a lifetime athlete and twenty-eight year raw fooder, is an advisor to world-class athletes and trainers from around the globe. He has worked professionally with top performers from almost every sport and field of entertainment, including such notables as tennis legend Martina Navratilova, NBA pro basketball player Ronnie Grandi son, track Olympic sprinter Doug Dickinson, pro women's soccer player Callie Withers, Chicken Soup for the Soul coauthor Mark Victor Hansen, and actress Demi Moore.

Dr. Graham is the author of several books on raw food and health, including The High Energy Diet Recipe Guide, Nutrition and Athletic Perform ance, and the forthcoming Prevention and Care of Athletic Injuries and The 80/10/10 Diet. He has shared his strategies for success with audiences at more than 4,000 presentations worldwide. Recognized as one of the fathers of the modern raw movement, Dr. Graham is the only lecturer to have attended and given keynote presentations at all of the major raw events in the world, from 1997 through 2005.

Dr. Graham is a founder of and is currently serving his third term as president of Healthful Living International, the world's premier Natural Hygiene organization. He is on the board of advisors of Voice for a Viable Future, the Vegetarian Union of North America, Living Light Films, and EarthSave International. He serves as nutrition advisor for the magazine Exercise, for Men Only and authors a column for Get Fresh! and Living Nutrition magazines.

Dr. Graham is the creator of "Simply Delicious" cuisine and director of Health & Fitness Weeks, which provide Olympic-class training and nutritional guidance to people of all fitness levels in beautiful settings around the world. He is living proof that eating whole, fresh, ripe, raw, organic food is the nutritional way to vibrant health and vitality

FoodnSport Can Help!

One of the most challenging aspects of changing your lifestyle is developing a support network. Often people are faced with unexpected challenges and are in need of experienced guidance that is beyond the scope of this material. Everyone at FoodnSport desires nothing more than to see you thriving successfully with this program. And we all know it's possible!

If you find that you still need personal direction or inspiration in living the 80/10/10 lifestyle after reading this book, you can take advantage of a number of options for one-on-one coaching or group learning experiences offered through my organization, FoodnSport. We offer personalized health/fitness/nutrition consultations for those interested in one-on-one guidance. In addition, FoodnSport offers one of a kind events annually including "Raw Health and Fitness Week", the Raw Nutritional Science Course, and Fasting Retreats in Costa Rica. Every event is designed to take your understanding, health and relationships to the next level.

A

All-American Pie 138
Almond Dip With Pineapple and Strawberries 134
Ambrosia 68
Any Nutty Icing 110
Apple-Walnut Grape Spread 106
Apricot Celery Dressing 122
Asparagus Soup 62
Avocado Icing 110

B

Banana Ice Cream 154
Banana Split 154
Berry Good Pie 142
Blackberry-Sesame Salad 82
Blue Cheeks 74
Borscht 50
Broccomole Salad 82

C

Cabbage Tomato Soup 44
Carobanana Mint Pie 138
Cashew Cream 108
Cashew Cucumber Dressing 112
Cashew Slaw 94
Caul Slaw 96
Cauliflower Soup 58
Celery Slaw 98
Celery-Red Pepper Soup 52
Chico Pudding 156
Citrus Ambrosia 66
Classic Sweet Shakes 26
Classic Tomato Celery Soup 50
Classic Tomato Salad 76
Classic "Juicy" Shakes 32
Coco Loco Dressing 118
Coconut Dip With Pear and Apple 134
Cookies 150
Cool Satsuma Dressing 112
Corn Salad 76
Could It Be Chocolate Pudding? 148
Creamy Corn Soup 56

Creamy Gazpacho Soup 54
Creamy Herb Dressing 122
Creamy Tomato Celery Soup 56
Crushed Berry Salad 80
Cuke Cashew Sandwiches 132
Cuke Soup 58

D
Dandy Candy 146

E
Eggless Salad 92

F
Figgy Pie 140
Figgy Puddin' 156
Five-Star Shake 30
Fruit Compote 70
Fruit Fondue 128
Fruit Gelleé 152
Fudge 148
Fudge S'Mores 150

G
Gazpacho 60
Gazpacho II 60
Glazed Strawberries 152
Good Ol' Ice Cream 156
Grape Soup 36
Grapefruit Soup 36
Greek Salad 78
Green Slaw 100
Green "Keys" Salad 90

H
Hawaiian Crush 104
Heirloom Avocado Salad 86
Herb Dressing 122
Holiday Pie 144

I
Italian Salad, Southern Style 78

J
Jam "Handwich" 70

K
Kiwi Cucumber Soup 42
Kiwi Orange Crush 72

L
Layered Salad 86

M
MacCabbage Salad 90
Mango Delight 74
Mango Fennel Soup 46
Mango Pie 142
Mango Red Pepper Dressing 114
Mango-Raspberry Dressing 116
Melon Soup 38

N
Nutty Fondue 128

O
Open Sesame Dressing 120
Orange and Walnut Sauce 106
Orange Fennel Slaw 100
Orange Juice Special Dressing 124
Orange Pepper Cuke Soup 54
Orange Verde Soup 42
Orange-Walnut Salad 84

P
Papaya Gazpacho 46
Papaya Salad 88
Papaya Spread 104
Peach Heirloom Tomato Soup 42
Peachberry Shake 28
Pineapple-Macadamia Dressing 120
Pineapple-Red Pepper Soup 40
Pistachio Cucumber Dressing 118
Piña Colada Sauce 104

Q
Quarter Shake 26

R
Rainbow Salad 92
Red Slaw 94

S
Salsa 108
Savory Sunrise Soup 48
Sprout Slaw 96
StrawberryMango Bliss 68

Strawberry Soup 36
Strawberry- Almond Salad 88
Strawberry-Cucumber Soup 40
Strawberry-Parsley 84
Sunflower Vegetable Dressing 114
Sunny's Favorite Dressing 118
Sweet Apple Soup 38
Sweet Grapes 72
Sweet Mango Pie 136
Sweet Russian Dressing 112
Sweet Tomato Walnut Dressing 124
Sweet Tomatoes 44

T

Tahini Dressing 116
Thick Shake 30
Tomato Basil Soup 52
Tomato Fennel Slaw 98
Tomato Heaven! 80
Tomato Olive Dressing 124
Tropical Delight 28
Tropical Lunch 66

V

Variation: Pineapple Rouge Dressing 114
Variation: Sweet Cukes 40
Vegetable Fondue 130
Veggie Nut Hors D'Oeuvres 132

Y

Your Basic Jam 108